Thalyson Vasconcelos Lima

Experimental Efficiency in Bean VCU Trials in Pernambuco

Thalyson Vasconcelos Lima

Experimental Efficiency in Bean VCU Trials in Pernambuco

ScienciaScripts

Imprint

Any brand names and product names mentioned in this book are subject to trademark, brand or patent protection and are trademarks or registered trademarks of their respective holders. The use of brand names, product names, common names, trade names, product descriptions etc. even without a particular marking in this work is in no way to be construed to mean that such names may be regarded as unrestricted in respect of trademark and brand protection legislation and could thus be used by anyone.

Cover image: www.ingimage.com

This book is a translation from the original published under ISBN 978-613-9-63548-1.

Publisher:
Sciencia Scripts
is a trademark of
Dodo Books Indian Ocean Ltd. and OmniScriptum S.R.L publishing group

120 High Road, East Finchley, London, N2 9ED, United Kingdom
Str. Armeneasca 28/1, office 1, Chisinau MD-2012, Republic of Moldova, Europe
Printed at: see last page
ISBN: 978-620-7-67655-2

"Your word is a lamp to my feet and a **light to my paths. " PSALMS 119:105**

To my parents Cleomar F. Lima and Mª Ivane de Melo, my brother Thalys Vasconcelos, for all their love and dedication to me, to my family for their affection and support at all times when I needed it.

I dedicate

ACKNOWLEDGEMENTS

To Jehovah God, who has been and always will be my guide, who has illuminated my path and that of my entire family, giving me the strength and courage to continue through the difficult moments of this arduous and long journey. To my parents, **Cleomar Ferreira Lima** and **Maria Ivane de Melo Vasconcelos Lima,** who are my safe harbour, for their example of life, perseverance and victory, for the years of dedication, for the nights spent awake, for their encouragement, unconditional and financial support, without which I would not be here today.

To my paternal grandparents **Elias Ferreira Lima** (in memoriam) and **Domingas Ferreira Lima** (in memoriam), my maternal grandparents **Vicente Ferreira de Vasconcelos** and **Maria Cardoso de Melo,** for their immense affection for me.

To my siblings, **Thalys Vasconcelos Lima** and **Maria de Fátima da Silva Moraes,** especially **Thalys,** who always helped me, encouraged me and was with me in every decision of my professional life.

To my uncles and family, especially **Josimar de Melo Vasconcelos** for the moments of relaxation, conversations, teachings, advice and unconditional support for my studies, for which I will be grateful for life.

I would also like to thank the Federal Rural University of Pernambuco (UFRPE) for the opportunity to study for a Master's degree in Agronomy - Plant Genetic Improvement and the National Council for Scientific and Technological Development (CNPq) for granting me the scholarship. To the **Agronomic Institute of Pernambuco (IPA)** and to researcher **Dr Antonio Félix da Costa** for his friendship, patience, availability and hours given to my learning, learning that I will carry with me for life, also for the availability of all the material that served as the basis for the preparation of this work in which his help was immeasurable.

To my supervisor **Prof Dr Jose Wilson da Silva,** for his guidance, patience and trust during the development of this work. For the teachings offered during our time together, which will serve me throughout my professional life. **Dr Emmanuelle Rodrigues Araújo** for her friendship, knowledge and constant encouragement, which contributed immensely to the success of this work.

To my great friends from the postgraduate course that I gained here, **Ricardo Valadares, Daurivane Rodrigues, Jackline Terto, Jamíle Medeiros, Roberta Rocha, Flávia Gomes, Talyta Magalhães, Sérgio Santana, Djayran Sobral, Tonny Cantarelly, Dulcineia Freitas, Gérsia Gonçalves and Edilton Cavalcanti** who contributed directly or indirectly to this work, for the great friendship and the bonds that will remain forever, for the contribution of my friends **Islan Diego Espindola de Carvalho and Carla Caroline Alves,** who supported me in carrying out this work, to whom I wish you all happiness and success in your personal and professional lives.

To all the teachers on the postgraduate programme in Plant Genetic Improvement, for their teaching and dedication, which played a major role in my academic life and contributed to my education. To all those who may not have been mentioned, but who certainly contributed immensely to the realisation of this research.

Thank you very much!

SUMMARY

SUMMARY

EXPERIMENTAL EFFICIENCY IN VCU TESTS IN THE STATE OF PERNAMBUCO

In Brazil, numerous field trials are conducted every year with the aim of recommending improved cultivars. These trials must comply with the requirements established by the Ministry of Agriculture, Livestock and Supply, the body that standardises these experiments and calls them Cultivation and Use Value (VCU). The VCU rules for common beans (Phaseolus vulgaris L.) stipulate that the experiments should be conducted in at least three locations, during the representative harvests of each region, for two years. Recent studies have pointed out the problems of considering only the value of the experimental coefficient of variation when assessing the precision of an experiment. Given the lack of studies in the area of experimental precision for the state of Pernambuco, the aim of this study was to evaluate a new strategy for assessing the experimental precision of trials of common bean lines in the state of Pernambuco, with the aim of improving the use of the statistical tools CV, F and Selective Accuracy for selecting new cultivars. Statistical evaluations will be used to determine the most appropriate number of repetitions, coefficients of variation, which made it possible to observe differences in the F test and selective accuracy due to its ability to inform about the efficiency of the value of the cultivar's genotype for classifying experimental precision. Recent work suggests that in order to achieve an accuracy target of 90 per cent, the F-test values associated with cultivar effects in the analysis of variance should be greater than 5.0. The use of the F statistic is an alternative to also take into account the level of genetic variability of the characters. This information could be used in the future when the Ministry of Agriculture, Livestock and Food Supply revises the VCU standards to guide cultivar breeders.

Key words: Phaseolus vulgaris L., evaluation, statistical parameters.

CHAPTER 1

GENERAL INTRODUCTION

1. INTRODUCTION

The common bean (Phaseolus vulgaris L.) is one of the most important and traditional agricultural crops in Brazil. Cultivated mainly by small family farmers, it has played an important social role as a generator of labour and income in the countryside. Traditionally consumed in Brazil, beans are one of the main sources of food for Brazilians. It is a food with exceptional nutritional value, as it is rich in B vitamins, fibre, carbohydrates and proteins that are complementary to those found in cereals, as well as being low in fat and low in cost (Borges 2007).

Beans are cultivated in different production systems, from subsistence farming to more modern agricultural systems in terms of production technologies. This wide diversity of environmental conditions in which the common bean is produced requires cultivars adapted to the different growing conditions (Embrapa/Cnpaf 2007).

In Brazil, numerous field trials are conducted every year with a view to recommending improved cultivars. In these trials, the requirements proposed by the Ministry of Agriculture, Livestock and Supply (MAPA) must be followed, which establishes a series of standards to determine the value of a new cultivar with a view to its registration, called Cultivation and Use Value (VCU). The VCU standards for beans lay down some basic requirements for a cultivar to be able to obtain the National Register of Cultivars (RNC): that experiments be conducted in at least three locations, in the representative crops of each region, over two agricultural years (Brasil 2007).

According to Vieira et al. (2006), various biotic and abiotic problems affect the bean crop and consequently also affect grain production, so these should be the objectives of breeding programmes. Once advanced lines have been developed, they should be tested in trials in different years and environments (Ramalho 2012, Cruz et al. 2014).

The genetic differences between strains to be detected in experiments are getting smaller and smaller. For this reason, alternatives must be sought to identify these differences (Kempton and Fox 1997). These include the number of repetitions, the size and format of the plots, the acceptable significance level for type I error and the use of environmental variables as covariates.

However, the common bean has a high level of genotype x environment interaction, which requires selected cultivars to be tested in different environments (locations, growing seasons and different years) in order to be more certain about their selection and recommendation.

Due to the importance of these trials, various studies have been carried out to establish statistical parameters for assessing the accuracy and efficiency of these experiments. To this end, genotype competition trials are carried out in different locations, times and regions of the country, and the best performing strains are registered and launched on the market, where they are best adapted. With the development of cultivars adapted to different environments by research institutions, it has become possible to produce beans all year round with productivity gains (Abreu et al. 1994).

To assess the precision of these experiments, among other parameters, most researchers use the coefficient of variation. Gomes (2000) considers coefficients of variation to be low when they are less than 10%; medium when they vary between 10% and 20%; high when they are between 20% and 30% and very high when they are greater than 30%. However, according to Resende (2007), a coefficient of variation of at least 20% and three repetitions are inadequate to provide information on the quality of trials whose purpose is to predict the VCU. The number of repetitions and the value of the F-test should be taken into account, so they are more suitable for this purpose and should be used to infer the selective accuracy and precision of these trials.

Recent work has discussed the advantages of using statistics such as heritability, coefficient of determination, number of repetitions, F-test value and accuracy together to assess the precision of experiments and cultivar launches (Cargnelutti Filho and Storck 2007, Resende and Duarte 2007).

Thus, among the various statistical parameters used in the evaluation of experiments and the lack of studies on experimental precision for the state of

Pernambuco, this study aims to evaluate a new strategy for assessing the experimental precision of trials of common bean lines in the state of Pernambuco, seeking to improve the use of statistical tools CV, F and Selective Accuracy for the selection of new cultivars.

2. THEORETICAL FRAMEWORK

2.1. Origin and botanical aspects of the bean plant

The Fabaceae family comprises 650 genera, encompassing around 18,000 species, distributed in the Caesalpinioideae, Faboideae and Mimosoideae subfamilies (Polhill et al., 1981). Among the 650 genera, the genus Phaseolus is represented by 55 species and only five are cultivated: Phaseolus acutifolius L., Phaseolus coccineus L., Phaseolus lunatus L., Phaseolus polyanthus Greenman and Phaseolus vulgaris L. The P. vulgaris L. species stands out for being the oldest and most widespread cultivated species in the world, and in some places it is considered an essential food (Singh 2001, Cronquist 1998).

The common bean (Phaseolus vulgaris L.) belongs to the class Dicotyledon, subclass Rosidae, order Fabales and family Fabaceae (Silva and Costa 2003). It is a diploid species, with $2n = 2x = 22$ chromosomes (Singh et al. 1991). The genus Phaseolus originated on the American continent (Singh 2001) and comprises 50 species (Debouck 1991, Zizumbo-villarreal et al. 2005). Of these, only five have been domesticated: common bean (P. vulgaris L.), lima bean (P. lunatus L.), yucote bean (P. coccineus L.), tepari bean (P. acutifolius A. Gray) (Gepts and Debouck 1991, Singh 2001), and P.dumosus (formerly P. polyanthus Greeman) (Gepts 2005).

The most widely accepted hypothesis regarding the origin of beans is that from a central area on the western slopes of the Andes in northern Peru and Ecuador, the beans were dispersed to the north (Colombia, Central America and Mexico) and south (Bolivia, Argentina and southern Peru), resulting in two Mesoamerican and Andean gene pools (Bitocchi et al. 2012).

The common bean has alternate trifoliate leaves; its inflorescence is a raceme that can be axillary or terminal. Plants are essentially classified according to their growth habit, and can be determinate or indeterminate. Growth habit is defined by the plant's stem growth and flowering habit, among other characteristics. Plants with a determinate growth habit tend to develop inflorescences at the apex of the stem and lateral branches, have a limited number of nodes and flower from the apex to the base of the plant (Silva 2005).

There are four main types of growth habit: type I, where the plant is determinate and bushy; type II, where the plant is indeterminate and has short internodes; type III, where the plant is indeterminate and has long internodes; and type IV, which is similar to type III, but where the plants are more fickle and have longer internodes (Santos and Gavilanes 2006).

The fruit is a kind of legume, called a pod, made up of two valves joined by two sutures. Beans have exalbuminous seeds, i.e. without albumen. The seed coat can vary in colour from black to white, beige, yellow, brown, red and pink, and can be uniformly coloured or have streaks, spots or dots; it can also be opaque, shiny or of intermediate brightness, with or without a halo (Debouck 1991, Santos and Gavilanes 2006, Silva 2005).

2.2. Socio-economic importance

The common bean is one of the most economically and nutritionally important agricultural products and is a staple food in both underdeveloped and developing countries in tropical and subtropical regions, particularly in the Americas and East and South Africa. It has a high protein content, ranging from 20 to 28%, as well as high carbohydrate, fibre and considerable iron content (Yokoyama et al. 1996).

Cultivated in more than 100 countries and according to FAO statistics for 2015, considering all genera and species of beans, average world production was 23.3 million tonnes, of which the seven main dry bean producing countries, which together account for around 52% of production, are: India (14%), Myanmar (13%), Brazil (11%), which is the third largest producer, followed by the USA (4%), China and Mexico (4%) and Tanzania (3%), respectively (Fao 2015).

In Brazil, around 3,385.6 thousand tonnes were produced in 2016, achieving an average yield of 1.127 Kg/ha in a total cultivated area of 3,040.0 thousand hectares. Total national bean consumption has varied between 3.3 and 3.6 million tonnes, due to domestic availability and market prices which induce consumers to buy more or less of the product, i.e. demand outstripping supply on the local market (Conab 2016).

In general, Brazilian production has supplied the domestic market, with the exception of black and white beans, which depend on imports to meet the demands of the consumer market. In 2008, 172,000 tonnes of black beans were imported, mainly from Argentina, China and Bolivia (Agrianual 2010).

Beans are an almost obligatory part of the Brazilian diet, as they are considered the main source of protein consumed on a daily basis and are a product of high socio-economic importance (Carneiro and Parré 2005). When compared to animal protein, beans are less expensive, contributing an average of 20 per cent of the daily protein requirement. The starch stored in the beans accounts for approximately 65% of their total composition, while cellulose and hemicellulose fibres account for around 7% (Bassinello 2007).

Brazil is among the world's largest producers of common beans, however, bean cultivation per area is still low and the probable causes are associated with pests and pathogens, low fertility, toxic elements in the soil, which leads to a reduction in the development of the root system, as well as poor seed physiological quality, the use of seeds without adequate sanitary treatment, which generates problems related to germination and the spread of diseases in the field (Souza et al. 2006, Vieira and Rava 2000).

2.3. Genetic improvement of the common bean

Given the magnitude of the crop, common bean breeding programmes have been carried out in Brazil by a number of research institutions, which have been supplying the market with new cultivars. The Federal University of Viçosa (UFV), the Federal University of Lavras (UFLA), Embrapa Rice and Beans, Embrapa Temperate

Climate, the Agricultural Research Company of the State of Minas Gerais (Epaming), the Agronomic Institute of Campinas (IAC), the Agronomic Institute of Paraná (IAPAR), as well as other institutions seeking to develop new cultivars suitable for each region, stand out in terms of bean breeding programmes.

In the state of Pernambuco, bean research began at the Instituto Agronômico de Pernambuco (IPA) in 1960, which was made possible by the agreement signed between IPA and SUDENE through the **"Food Crops" project, Other agreements were signed and** projects were funded by various institutions, but it was EMBRAPA, through the National Rice and Bean Research Centre (CNPAF), which maintained a certain regularity in funding bean research for a long time, especially in the costing line (Costa and Lopes 1999). This partnership between the institutions continues to this day.

In this way, it was possible to start a breeding programme for the crop, with the collection and introduction of varieties, and competition trials of these varieties, divided into four stages, concluding with hybridisation between the best strains (Silva 1962). These trials were the basis for the programme's research: it was ascertained which cultivars stood out as having the best yields and which were resistant to diseases (rust - the biggest phytosanitary problem at the time).

Based on these initial cultivars, a series of trials were started throughout the state of Pernambuco, covering 13 municipalities, from the Zona da Mata to the Chapada do Araripe, involving various areas such as insect control, planting time, density and spacing, mineral fertilisation, liming, cultivar competition and the use of inoculants (Sudene 1967).

As a result of this work, especially the cultivar competitions, it became possible to start the breeding programme itself, with the first crosses being made in 1966. Some of the cultivars created by the Pernambuco Agronomic Institute's bean breeding programme stand out: Ipa 74-19, Ipa-1, Ipa-3, Ipa-4, Ipa-5, Ipa-6, Ipa-7, Ipa-8, Ipa-9, BR-Ipa-10, BR-Ipa-11, Brígida and Princesa (Costa and Lopes 1999).

The world's largest collection of bean germplasm has more than 38,000 accessions and is located at the International Centre for Tropical Agriculture (CIAT) in Colombia. In Brazil, Embrapa Rice and Beans, together with Embrapa Genetic Resources and Biotechnology, has the largest collection with approximately 14,100 accessions. UFV also has an Active Germplasm Bank with an average of 600 accessions. This wide genetic variability of the bean plant makes it easier to choose suitable genotypes for genetic improvement of the crop in order to achieve the objectives of each programme (Celin 2011).

Improved cultivars combine desirable characteristics such as resistance to diseases, plant architecture that is more suited to the reality of each producer and greater productive potential, thus contributing to the increase in crop productivity, which was 729 kg/ha in 1997 to 1,160 kg/ha on average in 2009 (Feijão 2011). According to Matos (2005), the productivity gains obtained in the UFLA common bean genetic improvement programme between 1974 and 2004 are in the order of 1.6% per year. Although continuous gains have been made by the breeding programmes, as the level of productivity of the cultivars increases, it becomes more difficult to identify superior genotypes, since the genotypic differences to be detected become smaller and smaller.

Due to the crop's low average yield, improving the productive performance of common beans associated with desirable agronomic characteristics has received greater attention from genetic improvement programmes for this crop (Singh 2001). The common bean is an autogamous species with cross-fertilisation rates of less than 5%. Therefore, the methodologies used to evaluate the lines generated by breeding programmes seek to obtain homozygous lines with characteristics superior to those found in the crop.

Since the aim is to bring together favourable alleles present in various genotypes, bean improvement is mainly based on hybridising cultivars and/or lines in order to generate segregating populations, in which the best lines are selected, which require more efficient and viable strategies to guarantee the success of the improvement programmes (Zimmermann et al. 1996, Rocha et al. 2009).

2.4. Experimental precision and crop value and use (VCU) trials

In order to try and unravel the effect of the interaction between strains and environments, the main alternative is to set up experiments to evaluate strains in as many environments as possible. In this way, it is possible to identify the most adapted strains, with the highest averages and the most stable, i.e. those that always perform above average. Numerous studies have been carried out to identify strains/cultivars that are better adapted and more stable (Pereira et al. 2009, Rocha et al. 2010, Barili at al. 2015, Azevedo at al. 2015).

The Ministry of Agriculture, Livestock and Supply (MAPA) implemented rules for the registration of new cultivars in the country with the decree of the Cultivar Protection Law (Law No. 9.456/97), and there was a need to standardise the Cultivation and Use Value (VCU) trials for the registration of new bean cultivars with the National Cultivar Protection Service (SNPC). Fearful of the genotypes x environments interaction, the bean VCU standards recommend that experiments should be conducted in at least three locations, in the representative harvests of each region, over two years (Brasil 2007). A first question is whether this number of environments would be ideal in order to obtain good statistical inference in the actual prediction of genetic values and to estimate how many repetitions would be sufficient for safer selection of new cultivars.

Some statistical parameters have been used to check the precision and quality of agricultural trials. Traditionally, the main parameter used is the experimental coefficient of variation (CV), which must be at an acceptable level for the species and character being assessed. However, this has been questioned because it does not take into account the crop being studied and, above all, the character being considered (Ramalho et al. 2000). However, according to Resende (2007) and Cargnelutti Filho and Storck (2007), a coefficient of variation of at least 20% and a minimum of three repetitions are inadequate to certify the quality of trials whose purpose is to inform the cultivation and use value (VCU) of genotypes.

In addition to using the coefficient of variation, other parameters related to experimental precision have been proposed. These parameters also take into account the number of repetitions used in the experiment, as well as residual variation and the value of the F-test. They are therefore more suitable for this purpose and should be used to infer selective accuracy and precision in VCU trials (Resende 2002, Cargnelutti Filho and Storck 2007).

Work related to the establishment of critical values of CV, for quality control of trials, and new criteria or methods for classifying the precision of experiments, based on the CV of a set of similar trials, has been developed in various crops, types of experiments and characters (Clemente and Muniz 2002, Costa et al. 2002, Carvalho et al. 2003, Lima et al. 2004, Resende 2002, 2007, Cargnelutti Filho and Storck 2007).

In these trials of competing cultivars, care must be taken to maintain the experimental conditions as homogeneous as possible, and reducing experimental error is a common objective of all researchers, with the aim of improving experimental precision and, consequently, obtaining more accurate estimates of the mean or other parameters (Ramalho et al. 2000).

3. BIBLIOGRAPHICAL REFERENCES

Abreu AFB, Ramalho MAP, Andrade MJB, Pereira Filho IA (1994) Progress of bean genetic improvement in the 1970s and 1980s in the South and Alto Paranaíba regions of Minas Gerais. **Pesquisa Agropecuária Brasileira 29**: 105-112.

Agrianual (2010) **Yearbook of Brazilian Agriculture**. Beans. FNP Institute: São Paulo, 318-323.

Azevedo CVG, Ribeiro T, Silva DA, Carbonell SAM, Chiorato AF (2015) Adaptability, stability and resistance to pathogens in bean genotypes **Pesquisa agropecuária brasileira 50**: 912-922.

Barili LD, Vale NM, Amaral RC, Carneiro JES, Silva FF, Carneiro PCS (2015) Adaptability and stability and grain yield in black bean cultivars recommended in Brazil in the last five decades. **Ciência Rural 45**: 19801986.

Bassinello PZ (2007) Grain quality. Brasília: Embrapa Rice and Beans. Available at
:

<http://www.agencia.cnptia.embrapa.br/Agencia4/AG01/arvore/AG01_2_2810200
4161635>. Accessed on: 12 February 2016.

Bitocchi E, Nanni L, Bellucci E, Rossi M, Giardinni A, Zeuli OS, Papa R (2012) Mesoamerican origin of the common bean (Phaseolus vulgaris L.) is revealed by sequence data. **Proceedings of the National Academy of Sciences of the United States of America109:** 788-796.

Borges MHC (2007) **Agronomic evaluation, stability and adaptability of common bean genotypes.** In: (Dissertation), Universidade Federal de Uberlândia, Uberlândia - MG. 91p.

Brazil (2007) Ministry of Agriculture, Livestock and Supply. **Minimum requirements for determining the cultivation and use value of beans (Phaseolus vulgaris), for entry in the national register of cultivars -** RNC. Technical Report, Annex IV, Brasilia, 18p.

Cargnelutti Filho A and Storck L (2007) Statistics for evaluating experimental precision in trials of maize cultivars. **Pesquisa Agropecuária Brasileira 42:** 17-24.

Carneiro PT and Parré JP (2005) Importance of the retail sector in the commercialisation of beans in Paraná. **Revista de Economia e Agronegócio 3:** 277-298.

Carvalho CGP, Arias CAA, Toledo JFF, Almeida LA, Kiihl RAS, Oliveira MF, Hiromoto DM, Takeda C (2003) Proposed classification of coefficients of variation in relation to soybean yield and plant height. **Pesquisa Agropecuária Brasileira 38:** 187-193.

Celin EF (2011) **Morphoagronomic characterisation of accessions from UFV's active bean germplasm bank.** In: Dissertation (Master's Degree in Genetics and Breeding) - Federal University of Viçosa, Viçosa, 43p.

Clemente AL and Muniz JA (2002) Evaluation of the coefficient of variation in experiments with forage grasses. **Ciência e Agrotecnologia 26:** 197-203.

Conab (2016) Monitoring the Brazilian Grain Harvest, 2015/2016 Harvest - Fifth survey, Brasília, p. 1-182, February 2016. Available at <http://www.conab.gov.br/conabweb/download/safra/7levsafra.pdf>. Accessed on 07 March 2016.

Costa AF and Lopes LHO (1999) Genetic resources and improvement of the common bean in Pernambuco. In: Queiroz MA, Goedert CO, Ramos SRR. (Ed.). Genetic resources and plant breeding for the Brazilian Northeast. Petrolina: Embrapa Semi-Arido; Brasília, DF: **Embrapa Recursos Genéticos e Biotecnologia**, 220-230.

Costa NHAD, Seraphin, JC, Zimmermann FJP (2002) New method for classifying coefficients of variation for upland rice cultivation. **Pesquisa Agropecuária Brasileira 37**: 243-249.

Cronquist A (1998) Devolution and classification of flowering plants. **Botanical Garden, New York, 555p.**

Cruz CD, Carneiro PCS and Regazzi AJ (2014) **Biometric models applied to genetic improvement**. Editora UFV, Viçosa v.2, ed.3, 668p.

Debouck, DG (1991) Systematics and morphology. In: Schoonhoven AV, Voysest O (ed.) Common beans: **Research for crop improvement**. C.A.B. Intl., Wallingford, UK and CIAT, Cali, Colombia. 55-118.

Embrapa/Cnpaf (2007) Bean **cultivation**. Embrapa Rice and Beans. Available at:<http://www.cnpaf.embrapa.gov.br/pesquisa/feijão.html>. Accessed on 19 Feb. 2016.

Fao - Food and Agriculture Organisation. Statistic/FAOSTAT (2015) Available at:<http://www.agricultura.pr.gov.br/arquivos/File/deral/Prognosticos/2016/_feijao _2015_16.pdf> Accessed on: 27 March 2016.

Feijão (2011) Economic **data on bean (Phaseolus vulagaris L.) and cowpea (Vigna unguiculata (L.) Walp) production in Brazil**: 1985 to 2009. Available at:<http://www.cnpaf.embrapa.br/apps/socioeconomia/index.htm>. Accessed on: 11 December 2016.

Gepts P (2005) Management of germplasm and pre-breeding in common bean. VIII Congresso nacional de pesquisa de feijão. Goiânia, p. 1200-1210.

Gomes FP (2000) **Curso de estatística experimental**. Ed. 13, São Paulo: Nobel, 479p.

Kempton RA and Fox PN (1997) (Ed.). **Statistical methods for plant variety evaluation.** London: Chapman & Hall, 191p.

Lima LL, Nunes GHS, Bezerra Neto F (2004) Coefficients of variation of some characteristics of the melon tree: a proposal for classification. **Horticultura Brasileira 22**: 14-17.

Matos JW (2005) **Critical analysis of the UFLA common bean genetic improvement programme from 1974 to 2004.** In: Thesis (PhD in Genetics and Plant Breeding) - Federal University of Lavras, Lavras, MG, 116p.

Pereira HS, Melo LC, Faria LC, Peloso MJD, Costa JG, Rava CA, Wendland A (2009) Adaptability and stability of common bean genotypes with carioca type grains in the Central Region of Brazil. **Pesquisa agropecuária Brasileira 44**: 29-37.

Ramalho MAP, Abreu AFB, Santos JB, Nunes JAR (2012) **Applications of quantitative genetics in the improvement of autogamous plants.** Lavras-MG. Editora UFLA. 522p.

Ramalho MAP, Ferreira DF, Oliveira AC (2000) **Experimentation in genetics and plant breeding.** Editora UFLA, Lavras, MG, 326p.

Resende MDV (2002) Biometric and statistical genetics in the improvement of perennial plants. **Embrapa Florestas,** Informação Tecnológica, Brasília. 975p.

Resende MDV (2007) Mathematics and statistics in the analysis of experiments and in genetic improvement. **Embrapa Florestas,** Colombo. 435p.

Resende MDV and Duarte JB (2007) Precision and quality control in cultivar evaluation experiments. **Pesquisa Agropecuária Tropical 37**: 182-194.

Rocha F. et al. (2009) Selection of bean (Phaseolus vulgaris L.) mutant populations for adaptive traits. **Revista Biotemas 22**: 20-27.

Rocha VPC, Moda-Cirilo V, Destro D, Fonseca Júnior NS, Prete CEC (2010) Adaptability and stability of the grain yield trait of the carioca and black bean commercial groups. **Ciências Agrárias 31**: 39-54.

Santos JB and Gavilanes ML (2006) Botany. In: Vieira C, Paula Júnior TJ, Borém A. (ed.) **Feijão: aspectos gerais e cultura no Estado de Minas Gerais.** Viçosa: Editora UFV, p.

55-81.

Silva HT (2005) Minimum descriptors indicated for characterising cultivars/varieties of common bean (Phaseolus vulgaris L.). Santo Antônio de Goiás: **Embrapa-CNPAF**, 32p. (Documents, 184).

Silva HT and Costa AO (2003) Botanical characterisation of wild species of the genus Phaseolus L. (Leguminosae). Santo Antônio de Goiás, **Embrapa/CNPAF**: 40p. (Documents, 156).

Silva JF (1962) Improvement plan for common beans, mulatinho type. In: REUNIÃO DE INVESTIGAÇÃO AGRONOMICA DO NORDESTE, 2. 1962, Recife. Anais...Recife: SUDENE, p. 83-87.

Singh SP (2001) Broadening the genetic base of common bean cultivars: a review. **Crop Science 4**: 1659-1675.

Souza RF, Faquin V, Fernandes LA, Avila FW (2006) Phosphate nutrition and bean yield under the influence of liming and organic fertiliser. **Ciência e Agrotecnologia 30**: 656 -664.

Sudene (1967) Superintendence for the Development of the Northeast (Recife -PE) Contribution to the study of food plants. Recife: SUDENE - Documentation Division, (BRASIL. SUDENE. Food Cultures - Study of Pernambuco, 1.

Vieira C, Paula Júnior TJ and Borém A (2006) (eds.) **Feijão**. 2. ed. Viçosa, Editora UFV, 600p.

Vieira EHN and Rava CA (2000) Bean seeds: production and technology. Santo Antônio de Goiás: **Embrapa Arroz e Feijão**, 600p.

Yokoyama LP, Banno K, Dluthcoski J (1996) **Socioeconomic aspects of bean cultivation**. In: Araújo RS, Rava CA, Stone LF, Zimmermann MJO (eds.) Common bean cultivation in Brazil. Piracicaba: POTAFOS, p. 1-20.

Zimmermann MJO et al.(1996) Genetic improvement and cultivars. In: Araújo RS et al. (Ed.). **Common bean cultivation in Brazil**. Piracicaba: POTAFOS, p. 224273.

CHAPTER 2

STRATEGIES FOR EVALUATING EXPERIMENTAL PRECISION IN BEAN GROWING IN THE STATE OF PERNAMBUCO

Thalyson Vasconcelos Lima[1], Carla Caroline da Silva[1], Antonio Félix da Costa[2] Leonardo Cunha Melo[3], Helton Santos Pereira[3] and José Wilson da Silva[1].

[1]Federal Rural University of Pernambuco, Department of Agronomy, Postgraduate Programme in Agronomy - Plant Genetic Improvement, Dom Manoel de Medeiros, s/n, Dois Irmãos, CEP 52171-900 Recife, PE, Brazil. E-mail: thalysonagro@gmail.com, carlacarolinealves@hotmail.com, jwsamaral@hotmail.com[2] Instituto Agronômico de Pernambuco (IPA), Av. General San Martin, 1371, Bongi, CEP 50761-000, Recife, PE, Brazil. E-mail: felix.antonio@ipa.br[3] Embrapa Arroz e Feijão, km 12- Zona Rural, GO-462, CEP 75375-000, Santo Antônio de Goiás, GO, Brazil. E-mail: leonardo.melo@embrapa.br, helton.pereira@embrapa.br

SUMMARY

With the aim of evaluating selective accuracy and 14 other statistics as measures of the degree of experimental precision, grain yield data from 63 Crop Value and Use trials of bean cultivars (Phaseolus vulgaris L.) carried out in the state of Pernambuco by the Agronomic Institute of Pernambuco (IPA) in the 2009 to 2016 agricultural years were used. From the analyses of variance, the minimum, maximum, mean, standard deviation and coefficient of variation values were obtained for 14 statistical parameters for each trial, which were then correlated.

with the selective accuracy statistic. Based on the selective accuracy, the simple and relative frequencies of each trial were calculated for each class of experimental precision (low, moderate, high and very high). The repeatability coefficient and the number of repetitions needed to predict the true value of the genotypes were estimated. The selective accuracy statistic, repeatability coefficient and F-test value are more suitable than CV for jointly evaluating experimental precision in bean genotype competition trials. This study shows that 60.31% of bean cultivar competition trials have high precision.

Keywords: Phaseolus vulgaris L., experimental precision, selective accuracy.

INTRODUCTION

In Brazil, numerous field trials are conducted every year, known as Value for

Cultivation and Use (VCU), with the aim of recommending improved cultivars. These trials must follow the requirements established by the Ministry of Agriculture, Livestock and Supply, the body that standardises these experiments for the purposes of registering, recommending and protecting cultivars (Brasil 2007).

Due to the importance of these trials, several researchers have endeavoured to establish statistical parameters to assess the accuracy and efficiency of these experiments. To this end, competition trials between genotypes are carried out in different locations, times and regions of the country, and the best performing strains are registered and launched on the market, where they adapt best (Abreu et al. 1994).

To assess the precision of experiments, most researchers use the coefficient of variation (CV). Gomes (2000) considers coefficients of variation to be low when they are below 10%, medium when they are between 10% and 20%, high when they are between 20% and 30% and very high when they are above 30%. However, according to Resende (2007), a coefficient of variation of at least 20% and three repetitions are inadequate to provide information on the quality of trials whose purpose is to predict the cultivation and use value (VCU) of genetic materials.

Works such as Resende (2002) and Resende and Duarte (2007) discuss the use of selective accuracy and F-test value statistics for genotypes in order to obtain better experimental precision in trials. In this way, they demonstrate that in cultivar competition trials these statistics are associated with greater genetic variations and lower residual variances, and are more appropriate than using the coefficient of variation alone (Gomes 1990, Costa et al. 2002).

Some studies emphasise that the use of statistical parameters such as heritability, repeatability coefficient, F-test and accuracy to assess the precision of experiments and cultivar launches are more appropriate (Cargnelutti Filho and Storck 2007, Resende 2007). According to Resende (2002), selective accuracy has the property of ranking cultivars for selection purposes and also of inferring the genotypic value of cultivars in VCU trials.

Another parameter to be taken into account in the relative performance of the cultivars evaluated is the use of the coefficient of repeatability of grain yield, i.e.

estimating the probability that the expression of the character will be repeated in future evaluations (Cargnelutti Filho et al. 2009b). Thus, repeatability expresses the maximum value that heritability can reach with high values of this coefficient for any trait, and indicates that it is possible to predict the real value of individuals based on a given number of measurements (Martuscello et al. 2007).

Considering the number of repetitions used in the experiment, repeatability is the parameter that will make it possible to define criteria for discarding experiments to evaluate and recommend cultivars, and it should be associated with other parameters to make the recommendation of a cultivar more reliable (Gurgel et al. 2013). According to Resende (2007), this association should be with the coefficient of variation and the value of the F test for the effects of cultivars, which are more suitable for this purpose and should be used to infer selective accuracy and precision in trials of cultivation and use value.

Among the various strategies used to improve trials, there are no reports on their use in breeding programmes for the state of Pernambuco.

The aim of this work is to evaluate a new strategy for assessing the experimental precision of bean line trials in the state of Pernambuco.

MATERIAL AND METHODS

In all the experiments, the treatments consisted of 11 to 37 common bean strains, in experimental trials of cultivation and use value, where all the experiments contained witnesses, which were cultivars recommended for the state of Pernambuco. These experiments are part of the network of trials conducted in the state of Pernambuco by the Agronomic Institute of Pernambuco (IPA) between 2009 and 2016.

Grain yield data from 63 trials conducted in the 2009/2010, 2011/2012, 2013/2014 and 2015/2016 biennia were used, which were carried out in different municipalities in the state of Pernambuco.

All the experiments were conducted using a randomised block design, with three replications and plots of four rows measuring 4m, spaced 0.5 x 0.2m apart. Grain yield data was obtained from the two centre rows, with the useful area being 4m .^{2}

An analysis of variance (ANOVA) was carried out on each trial and the mean square of the block (QMB), MEAN SQUARE OF THE treatment (QMT), mean square of the error (QME), F-test value for genotype (F), overall mean of the trial (M), experimental coefficient of variation (CV), genetic coefficient of variation (CVg), heritability (h^2) and the ratio CVr=100 CVg/ CVe were recorded, according to Cruz (2008).

Experimental precision was verified using selective accuracy (SA), which corresponds to the linear correlation between genotypic and phenotypic values, where F is the value of the F test for genotype according to Resende (2002), together with the experimental coefficient of variation (CV), the minimum significant difference by the Tukey test at 5% probability, calculated using the formulae: Selective accuracy: AS= $(1-1/F)^{1/2}$; Experimental coefficient of variation: $CV(\%) = \frac{\sqrt{QMe}}{\bar{x}} . 100$; Tukey test: $\Delta(5\%) = q\frac{s}{\sqrt{r}}$; F test: $F = \frac{QMT}{QMR}$

Statistics such as: QMB, QME, QMT, M, CV, CVg, CVr, R^2 , h^2 , F, D, A and A/D were obtained for each of the 63 trials, and the minimum and maximum values, mean and dispersion measure (CV) were calculated using the Lilliefors normality test (Campos, 1983). The evaluations were also carried out by estimating the repeatability coefficient (k) of each trial, estimated by three statistical procedures using the following methods: analysis of variance (ANOVA), principal components and structural analysis. The minimum number of repetitions estimated to predict the real value of the individuals based on the pre-established 0.81 coefficient of genotypic determination (R2) was calculated according to the expression η=[R^2 (1- k)]/[1- R2) k] provided by Cruz et al. (2012).

From (k), which is the average of the repeatability coefficients of the 63 trials, the coefficient of determination was calculated as a function of the number of repetitions (η=3), **to graphically show the behaviour of the relationship between R^2 and η. The R^2 , which represents the certainty of predicting the real value of the selected genotypes, based on q measurements taken, was obtained from the expression $R^2 = (J -k)/[1 + -k (J -1)]$** (Cruz 2008).

Subsequently, Pearson's linear correlation coefficient (r) was calculated between AS and the statistics Mean, QMB, QMe, QMT, CV, CVr, CVg, R^2 , F, h^2 , D, A and A/D, and their significance was verified using Student's t-test at 5% probability.

Subsequently, on the basis of AS and F, the simple and relative frequencies of trials were calculated, established in classes of experimental precision in the low, moderate, high and very high ranges according to Resende and Duarte (2007). The statistical analyses were carried out using the GENES programme (Cruz 2008) and the experimental precision parameters were calculated using the Office Excel software.

RESULTS AND DISCUSSION

In general, the frequencies presented in all the trials were classified as low (19.04%), moderate (20.63%), high (44.44%) and very high (15.88%) experimental precision (Figure 1). Adding up the frequencies of trials with low selective accuracy (AS < 0.5) and those **with moderate accuracy (> 0.50 and ≤ 0.70) brings the total to** around 39.7%, so that **60.3% of trials have high accuracy (> 0.70 and ≤ 0.90) and very high accuracy (> 0.90). When the frequencies of tests in the** high and very high precision **ranges are added together**, the tests carried out in the 2009/2010 and 2013/2014 biennia are superior, reaching 30.15%, totalling 19 tests with excellent experimental precision for these two biennia (Table 1).

It can be seen that in order to achieve an ideal selective accuracy of 90% or more, corresponding to coefficient of determination values above 80%, as recommended by Storck et al. (2000), for a safe statistical inference the F values for cultivars must be equal to or greater than 5, while in this study around 10% of the trials showed higher values for the F test, which showed very high experimental precision (AS= > 0.90).As such, this can be a reference value for experiments evaluating experimental precision, as it is independent of the species grown and the character being evaluated, and can serve as a standard for evaluating experiments of this nature.

Table 2 shows the correlation values between selective accuracy (SA) and the other statistics evaluated in bean cultivation and use value trials. All correlations were significant at 5% probability using the t-test.

Among the significant correlations, positive ones were found between AS and the variables Q_{MT} (0.36*), CVg (0.80*), CVe (0.79*), R^2 (0.94*), F (0.53*), h^2 (0.96*), A (0.12*) and AD (0.61*). Indicating that the higher the values found for AS, the higher the values of the correlated variables. CVg, CVe, R2 and h2 were higher than 79%, providing greater efficiency in indirect selection. The variables F and AD showed medium relationships and A and QMt low relationships, thus providing lower gains.

This study also found negative correlations between (AS) and the variables Mean (-0.04*), QMb (-0.05*), QMe (-0.26*), CVe (-0.29*) and D (0.38*). These results reveal an inversely proportional relationship, indicating that these statistics are associated with greater environmental variance and are independent of the genetic variability of the genotypes evaluated.

According to Gonçalves et al. (2008), correlations are appropriate for evaluating the association between characteristics because they are dimensionless and allow comparisons to be made between different pairs of characteristics.

Reinforcing the use of the correlation between the AS statistic and the other parameters to obtain experimental precision, it can be seen that most of these statistics are positively correlated, and are linked to greater genetic variability and lower residual variances, thus being suitable for classifying the precision of experiments, according to Cargnelutti Filho and Storck (2007).

Therefore, for competition trials of superior genotypes, it is important that the experimental precision statistic takes genetic variability into account (Resende and Duarte 2007) and is independent of the mean. Thus, the CV and DMS statistics, traditionally used for this purpose, can be replaced by other statistics, such as AS and F, which are considered more appropriate (Resende 2002, Cargnelutti Filho and Storck 2007, 2009, Resende and Duarte 2007, Cargnelutti Filho et al. 2009b).

In view of this, the selective accuracy statistic is suitable for assessing the experimental precision of bean genotype cultivation and use value trials, and discarding trials based solely on the coefficient of variation is inadvisable, as seen in Figure 2, which, despite having a high CV (37.2 per cent), obtained high experimental precision (AS= 0.80), taking into account the criteria for classification, such as selective accuracy.

Separating the eight trials according to the low AS criteria (Table 3), it can be seen that in this group there is consistently no difference between the genotypes.

However, the average CV of this group (21.8%) is relatively close to the overall average CV (19.7%). According to the SA criterion, these trials have low precision, as indeed they do, since the average D value for this group (D =1.5) is higher than the overall average (D =0.90).

Using the criterion of discarding 5% of the trials with the highest CVs, i.e. $P(CV>c) = 0.95$, the normal distribution shows that **"m" and "dp" are the** mean and standard deviation of the CV values (Storck et al. 2016). In this case, as m=19.72 and dp= 5.54 (Table 2) we have c=28.72 and $P(CV>28.72)=0.95$. Of the seven trials that fell into this $P(CV>28.8)=0.95$ category (Table 4), i.e. trials that showed high values for the CV criterion, three of these would be discarded because they had low precision for the AS criterion and showed very high CV.

The selective accuracy values presented in this study show that **even with some trials with a high CV = ≥ 20% (Figure 2), good experimental precision was obtained (AS = ≥ 0.75), making it possible to select the** best genotypes for the environments evaluated and subsequently launch new bean cultivars on the market. According to Storck et al. (2010), the AS statistic depends not only on the magnitude of the experimental error and the number of repetitions, but also on the proportion between the genetic and residual variations associated with the character being evaluated.

Based on the bean grain yield, the average repeatability coefficient of the 63 trials was obtained (k=0.35), which indicates that **eight repetitions** would be necessary, i.e. η = **0.81**(1 - 0.35)/(1 - 0.81)0.35 = 7.9, to affirm, with 81% accuracy ($R^2 = 0.81$), the superiority of a given genotype for the conditions of the state of Pernambuco (Figure 3).

According to Resende and Duarte (2007), trials with R2 equal to or greater than 81%, i.e. with selective accuracy equal to or greater than 90%, should be aimed for, as they would constitute trials with very high experimental precision. However, in practice it is not possible to conduct trials with eight repetitions. Thus, when **the reverse calculation is carried out, starting with H= 8 repetitions, the estimated value of the** genotypic **coefficient of** determination is R2 = (8 x 0.35)/[1 + 0.35(8 - 1)] = 0.81, while for $\eta = 3$ **the R^2 is 0.61, i.e., the coefficients of determination (R^2) associated with the estimates of the** repeatability coefficients (k) were equal to or greater than 0.61, regardless of the estimation method for the trials, indicating that three repetitions make it possible to select genotypes with 61% reliability in predicting the genotype's true value.

When considering the repeatability coefficient values for each experimental precision range, we have (k)= 0.17; 0.42; 0.66 and 0.35 for low+moderate, high, very high and the general average of experimental precision respectively. In order to infer possible superior genotypes in terms of grain yield with 81% precision, trials with k= 0.17 would require 20 repetitions, while trials with k= 0.42 and k= 0.66 would require 6 and 3 repetitions respectively (Table 5).

In order to evaluate grain yield in trials that make it possible to identify superior bean genotypes in terms of grain yield with 81% accuracy in predicting their real value, eight repetitions would be necessary. In practice, particularly in these groups of bean trials carried out in the state of Pernambuco, selective accuracy targets of 90 per cent, which is equivalent to a genotypic coefficient of determination of 81 per cent, presented a greater number of repetitions, theoretically, than the six and four repetitions recommended for beans, found by Resende and Duarte (2007) and Cargnelutti Filho et al. (2009a). However, the use of a greater number of repetitions should be encouraged in order to maximise experimental precision.

CONCLUSIONS

The selective accuracy statistic and the value of the F-test prove to be advantageous in relation to the coefficient of variation for assessing the experimental precision of bean genotype competition trials; using this statistic, it can be said that 60.3 per cent of the trials have high or very high experimental precision.

ACKNOWLEDGEMENTS

To the National Council for Scientific and Technological Development (CNPq) for the postgraduate scholarship.

BIBLIOGRAPHICAL REFERENCES

Abreu AFB, Ramalho MAP, Andrade MJB, Pereira Filho IA (1994) Progress of bean genetic improvement in the 1970s and 1980s in the South and Alto Paranaíba regions of Minas Gerais. **Pesquisa Agropecuária Brasileira 29**: 105-112.

Brazil (2007) Ministry of Agriculture, Livestock and Supply. National Register of Cultivars (RNC) - Technical Report. Minimum requirements for determining the value of cultivation and use, for registration in the RNC. Brasília, p. 18.

Campos H (1983) **Non-parametric experimental statistics**. 4.ed. Esalq, Department of Mathematics and Statistics, Piracicaba, 349p.

Cargnelutti Filho A, Ribeiro ND, Storck L (2009a) Number of replications for comparing bean cultivars. **Ciência Rural 39**: 2419-2424.

Cargnelutti Filho A and Storck L (2007) Statistics for evaluating experimental precision in trials of maize cultivars. **Pesquisa agropecuária brasileira 42**: 17-24.

Cargnelutti Filho A, Storck L, Ribeiro ND (2009b) Measures of experimental precision in trials with bean and soya genotypes. **Pesquisa Agropecuária Brasileira 44**: 1225-1231.

Cargnelutti Filho A and Storck L (2009) Measures of the degree of experimental precision

in competition trials of maize cultivars. **Pesquisa Agropecuária Brasileira 44**:111-117.

Costa NHAD, Seraphin JC, Zimmermann FJP (2002) New method for classifying coefficients of variation for upland rice cultivation. **Pesquisa Agropecuária Brasileira 37**: 243-249.

Cruz CD, Regazzi AJ, Carneiro PCS (2012) **Biometric models applied to genetic improvement**. Viçosa, UFV, 508p.

Cruz CD (2008) **GENES programme**: computational application in genetics and statistics UFV, Viçosa, MG. 278p.

Gomes PF (2000) **Curso de estatística experimental**. Ed. 13, São Paulo, Nobel, 479p.

Gonçalves GM, Viana AP, Reis LS, Neto FVB, Amaral Júnior AT, Reis LS (2008) Phenotypic and genetic-additive correlations in yellow passion fruit using design I. **Revista Ciência e Agrotecnologia 32**: 1413-1418.

Gurgel FL, Ferreira DF, Soares ACS (2013) The coefficient of variation as an evaluation criterion in maize and bean experiments. **Embrapa Amazônia Oriental**, Belém, 80p.

Martuscello JA, Jank L, Fonseca DM, Cruz CD, Cunha DNFV (2007) Repeatability of agronomic characters in Panicum maximum Jacq. **Revista Brasileira de Zootecnia 36**: 1975-1981.

Miranda JEC, Costa CP, Cruz CD (1988) Genotypic, phenotypic and environmental correlations between fruit and plant characters in chilli (Capsicum annuum L.). **Revista Brasileira de Genética 11**: 457-468.

Resende MDV (2002) Biometric and statistical genetics in the improvement of perennial plants. **Embrapa Informação Tecnológica**, Brasília, 975p.

Resende MDV (2007) Mathematics and statistics in analysing experiments and genetic improvement. **Embrapa Florestas**, Colombo, 435p.

Resende MDV and Duarte JB (2007) Precision and quality control in cultivar evaluation experiments. **Pesquisa Agropecuária Tropical 37**: 182-194.

Storck L, Cargnelutti Filho A, Lúcio AD, Missio EL, Rubin SAL (2010) Evaluation of experimental precision in competition trials of soya cultivars. **Pesquisa Ciência e Agrotecnologia 34**: 572-578.

Storck L, Garcia DC, Lopes SJ, Estefanel V (2016) **Plant experimentation**. 1ª reprint, 3.ed. Santa Maria: UFSM, 198p.

Storck L, Lopes SJ, Marques DG, Tisott CA, Ros CA (2000) Analysis of covariance to improve discrimination capacity in maize cultivar trials. **Pesquisa Agropecuária Brasileira 35**: 1311-1316.

Table 1. Frequency of 63 bean (Phaseolus vulgaris L.) Cultivation and Use Value (VCU) trials, in terms of the range of experimental precision according to selective accuracy conducted in the state of Pernambuco.

	Selective Accuracy Ranges								TOTAL
	Low		Moderate		High		Very high		
	$\leq 0,5$		$0,5 \leq 0,7$		$0,7 \leq 0,9$		$\geq 0,9$		
	Fi	Fr(%)	Fi	Fr(%)	Fi	Fr(%)	Fi	Fr(%)	
GENERAL	12	19,04	13	20,63	28	44,44	10	15,88	63
Year 09/10	0	0,0	1	11,11	7	77,77	1	11,11	9

Year 11/12	4	25	5	31,25	6	37,5	1	6,25	16
Year 13/14	4	23,52	2	11,76	5	29,41	6	35,29	17
Year 15/16	4	19,04	5	23,80	10	47,61	2	9,52	21

Low - (<0.5); Moderate - (>0.5<0.7); High - (>0.7<0.9); Very High - (>0.9); Fi - simple frequency; Fr(%) - relative frequency; AS - selective accuracy.

Table 2. Linear correlation coefficient (r) between selective accuracy (SA), maximum, minimum, mean, standard deviation (SD), and coefficient of variation (CV) based on 63 Crop Value and Use (VCU) trials of beans (Phaseolus vulgaris L.), in relation to grain yield.

Statistics	r	Minimum	Maximum	Average	dp	CV
AS	-	0	0,9799	0,7512	0,2694	35,8
Average	-0,0401*	225,2917	3132,679	1619,931	509,4158	31,4
QMb	-0,0525*	3844,313	1047698,3	208355,8	733765,1	352,1
QMe	-0,2614*	2911,538	505096,4	80048,45	83928,25	104,8
QMt	0,36623*	7273,525	2070582	180490,8	254097,2	140,7
CVe	-0,2983*	8,3647	43,2678	19,7255	5,5428	28,1
CVg	0,8048*	0	44,603	14,5795	7,3703	50,5
CVr	0,7984*	0	283,7474	65,7110	43,3183	65,9
R^2	0,9484*	0,3231	0,9617	0,6965	0,1274	18,2
F	0,5317*	0,4774	25,1537	2,2953	2,5977	113,1
Heritability (h2)	0,9666*	0	96,0244	57,8411	25,0393	43,2
D	-0,3882*	63,8510	1081,047	357,652	160,6854	44,9
Amplitude (A)	0,1245*	332,5	3842,5	1707,5	666,9815	39,0
AD= A/D	0,6180*	0,8272	11,3186	4,5881	1,2093	26,3

AS: selective accuracy; Mean: mean of all trials; QMb: MEAN square of block; QMe: MEAN SQUARE OF error; QMt: MEAN SQUARE OF treatments; CVe: coefficient of experimental variation; CVg: coefficient of genotypic variance; CVr: variation100.CVe/CVg; R2 = coefficient of determination; F: F test value for genotypes; h2: heritability; D: minimum significant difference between trials, by Tukey's test at 5% probability; A: range of trial means; r: linear correlation coefficient; Min: minimum trial values; Max: maximum trial values; Mean: mean trial values; sd: standard deviation; CV: coefficient of variation in percentage.

*Significant by t-test at 5% probability.

Table 3. Selective accuracy (SA), average grain yields (kg/ha), experimental coefficient of variation (CV), minimum significant difference by Tukey's test (D) and hypothesis adopted from the F test for differences between genotypes (H0 and H1) for bean trials with low experimental accuracy AS<0.5.

TESTS	AS	AVERAGE	CV	D	HYPOTHESIS
9	0,5	1319,16	21,44	0,826	Ho
18	0,2	657,22	33,24	0,661	H0
25	0,5	2084,92	25,07	1,637	H0
31	0,04	740,81	19,08	0,417	H0
41	0,3	1982,72	18,96	1,109	H0
45	0,2	3132,67	22,20	2,093	H0
49	0,5	1667,95	19,72	0,971	H1
51	0,2	2185,73	24,13	1,556	H0
Average of the 8	0,3	1,82	21,82	1,5	
Average of 63	0,75	1,62	19,72	0,90	

AS: selective accuracy; Mean: mean of each trial; CV: coefficient of experimental variation; D: minimum significant difference between trials, by Tukey's test at 5% probability; Hypothesis: Hypothesis test.

Table 4. Selective accuracy (SA), average grain yields (kg/ha), experimental coefficient of variation (CV), minimum significant difference by Tukey's test (D) and hypothesis adopted from the F test for difference between genotypes (H0 and H1) for the bean trials as a criterion for discarding P(CV>28.8).

TESTS	AS	AVERAGE	CVe	D	HYPOTHESIS
7	0,72	1,002	36,73	1,143	Ho
18	0,25	657,22	33,24	0,661	H1
28	0,0	1,538	38,32	1,774	H1
55	0,0	299,29	36,09	0,335	H1
57	0,68	1619,93	34,15	1,644	H1
62	0,70	2074,75	31,51	2,002	H1
63	0,87	251,31	30,82	0,237	H0
Average of the 7	0,46	1,2	34,15	1,7	

AS: selective accuracy; Mean: mean of each trial; CV: coefficient of experimental variation; D: minimum significant difference between trials, by Tukey's test at 5% probability; Hypothesis: Hypothesis test.

Table 5: Estimated number of repetitions (q), average coefficient of determination (R^2) of the trials in terms of experimental class for 63 Crop and Use Value trials for the bean crop (Phaseolus vulgaris L) in the state of Pernambuco.

Experimental Class	k- medium	R^2 - estimated	η
General	**0,35**	**0,81**	**7,79**
Low + Moderate	0,17	0,81	20
High	0,42	0,81	5,9
Very high	0,66	0,81	2,19

General: all tests; Low+moderate: average SA= < 0.5 <0.7; High: average SA= 0.7<0.9; very high: average SA= s 0.9; k: repeatability coefficient R^2 : coefficient of determination; q: number of repetitions.

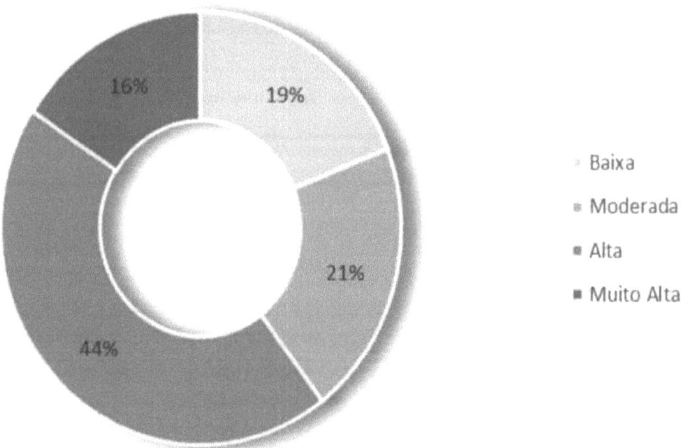

Figure 1. Graph of the percentage distribution of the 63 bean trials in terms of experimental precision classes in the state of Pernambuco.

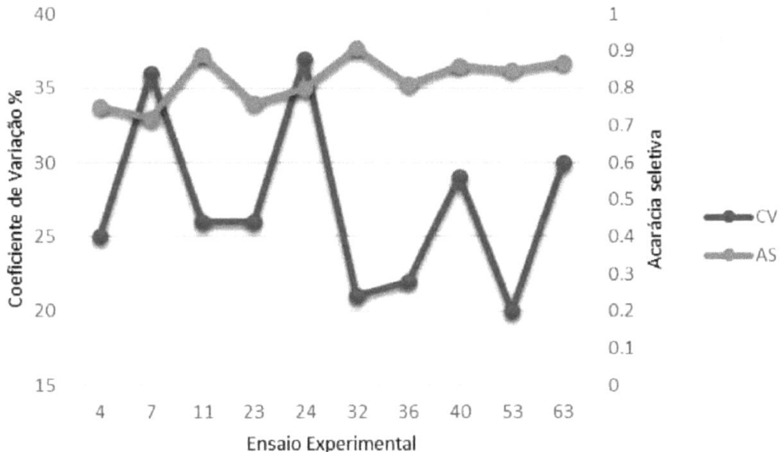

Figure 2: Values of selective accuracy (SA), coefficient of experimental variation (CV) for Crop Value and Use trials of beans (Phaseolus vulgaris L.) with high and very high experimental precision in the state of Pernambuco.

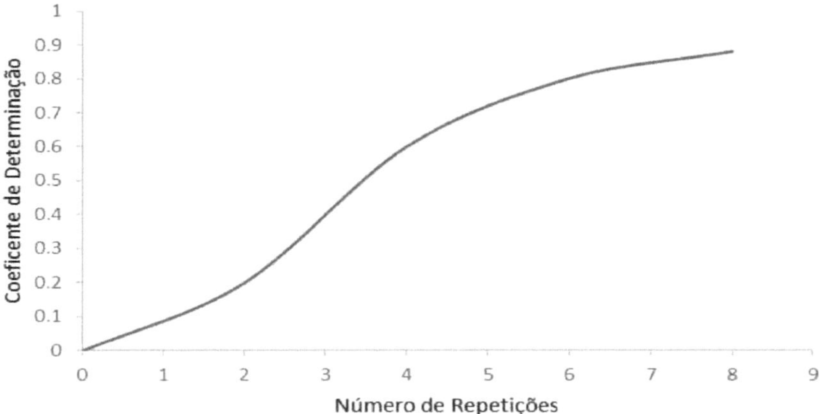

Figure 3 Number of repetitions ($r^\wedge$) associated with the coefficient of determination (R^2) estimated for the yield character in experimental trials of beans (Phaseolus vulgaris L.) in the state of Pernambuco.

CHAPTER 3

CBAB - CROP BREEDING AND APPLIED BIOTECHNOLOGY
Instructions for authors
General policy and scope of the journal
The **CBAB - CROP BREEDING AND APPLIED BIOTECHNOLOGY** (ISSN 1984-7033, on line version) - is the official quarterly journal of the Brazilian Society of Plant Breeding (www.sbmp.org.br), abbreviated CROP BREED APPL BIOTECHNOL. It is indexed in ISI Thomson Reuters, Scopus, AGRIS, CAB International Abstracts, Biosys, Latindex, Periódica, Chemical AbstractsService, Agricola, Agrobase, Wilson, Ebsco, DOAJ, Acervo Documental da Embrapa and Portal da Capes. It publishes original scientific articles which contribute to the scientific and technological development of plant breeding and agriculture. Articles should be to do with basic and applied research on improvement of perennial and annual plants, within the fields of genetics, conservation of germplasm, biotechnology, genomics, cytogenetics, experimental statistics, seeds, food quality, biotic and abiotic stress, and correlated areas. The article must be unpublished. Simultaneous submission to another periodical is ruled out. Authors are held solely responsible for the opinions and ideas expressed, which do not necessarily reflect the view of the Editorial board. However, the Editorial board reserves the right to suggest or ask for any modifications required. The journal adopts the Ithenticate software for identification of plagiarism. Complete or partial reproduction of articles is permitted, provided the source is cited. All content of the journal, except where identified, is licensed under a Creative Commons attribution-type BY. All articles are published free of charge. This is an open access journal.
Article
The **CBAB** publishes exclusively in English. The authors must submit their articles in English. It is mandatory that the article, after approval, be reviewed in linguistic terms, which must be made exclusively by the journal's official translators, the onus of this service being the responsibility of the author. Contributions are submitted via WEB, access **http://www.sbmp.org.br/cbab/siscbab/index.php** clicking **Submissi on**, whereupon the article registration system will automatically ask for a password and author's e-mail. **Delete all author and correspondence information from the manuscript file**. As the Journal has a blind review policy, authors should not reveal their identities in the manuscript. The author will be asked to enter this information in a separate form, during the submission process before uploading the manuscript file. The author can monitor the manuscript's stages of proceeding by his/her e-mail and personal password. Expert *ad hoc* reviewers evaluate the manuscripts to assist the Editorial Board with the final decision of approval, modification, or disapproval. The complete manuscript

should comply with the following sequence: title, abstract, key words, introduction, material and methods, results and discussion, acknowledgements, references, and tables and figures. The manuscript must be typed in Word for Windows, in times new roman 12, double spacing, A4 format, with 20 mm margins and consecutive top right numbering. The double spaced text must not exceed 18 pages, including separately placed tables and figures (one a page) at the end of the text. All the equations, models and symbols should be made in Microsoft Equation. The title should be clear, concise, and express the gist of the article. It should not exceed 15 words, be typed in bold, left, with initial upper case letters. The authors' full names, and their institutional addresses should be entered in the proof read. The abstract should not contain more than 150 words. A maximum of 5 key words, different from the title, are allowed. The introduction should include a brief literature review on subject and aims of the study. Material and Methods must enable other researchers to repeat the experience. Preferably, Results and Discussion should be presented together for ease of reading. Acknowledgements should be succinct, and limited to effective co-workers and funding agencies.Be careful about the references. Never cite summaries of events and theses, or any other unpublished literature. These measures will help shape a manuscript that will be a credit both to your article and to the journal. Citations mentioned in the text by the last name of the author and the year (for instance, Liu 1998, Pereira and Amaral Júnior 2001, William et al. 1990) are to be alphabetically listed in the References item, according to the following examples:

Articles in periodicals:
Pereira MG and Amaral Júnior AT (2001) Estimation of genetic components in popcorn based on the nested design. **Crop Breeding and Applied Biotechnology 1**: 3-10.

Book:
Ramalho MAP, Ferreira DF and Oliveira AC (2000) **Experimentação em genética e melhoramento de plantas**. Editora UFLA, Lavras, 326p.

Chapter of book:
Sakiyama NS, Pereira AA and Zambolim L (1999) Improvement of Arabica coffee. In Borém A (ed.) **Melhoramento de espécies cultivadas**. Editora UFV, Viçosa, p. 189-204.

Congress:
Frey KJ (1992) Plant breeding perspectives for the 1990s. In Stalker HT and Murphy JP (eds.) **Proceedings of the Symposium on Plant Breeding in the 1990s**. CAB, Wallingford, p. 113.

The **CBAB** publishes, besides articles, other text forms, equally subjected to the discretion of ad hoc reviewers.

Review
Leading authors of certain topics will be asked for a Review by the Editorial board (also restricted to 18 pages), which should shed light specifically on

stirring subject matters that deserve a deeper analysis into their stage of development.

Notes

Notes are limited to 12 pages and designated to inform about new studies or observations, wherefore the analytical tools are not required. They may focus on a matter of broad interest; briefly describe an original study; report on participatory research; express
observations of special interest in the fields of research, teaching, and applied Sciences; or comment on the release of new software in a plant breeding-related area.

Plant breeding programmes

Outstanding breeding programmes
regarding innovation, efficiency, impact, and/or continuity can be portrayed in the **CBAB**, restricted to 18 pages.

Cultivar release

New cultivars deserve special attention for their key role in plant breeding, and consequently, for domestic agriculture. A contribution to this section should comprise an abstract of maximally 50 words, key words, an introduction, mention the applied improvement methods, performance characteristics, foundation seed production, and contain a minimum of references (follow examples of articles references), tables, and figures. The entire text should not exceed 12 pages.

Book review

This new section was created to announce new books related to plant breeding. A contribution to this section should comprise two copies of the book sent in by the Author. The book will be revised by an expertise referee chosen by the Editorial Board to edit a brief.

Viewpoint

At the invitation of the Editorial board, viewpoints will be - as reviews - worked out for the **CBAB**, to outline topics which interest plant breeders and society.

Letters

Short letters of general interest are also welcome for publication, subject to changes by the Editorial Board for reasons of space limits or clarity of expression.

CBAB - CROP BREEDING AND APPLIED BIOTECHNOLOGY
Instructions to Authors

General policy and scope of the journal **CBAB - CROP BREEDING AND APPLIED BIOTECHNOLOGY** (ISSN 1984-7033, online version) - is the official quarterly journal of the Brazilian Society for Plant Breeding (http://www.sbmp.org.br). Its abbreviated international name is CROP BREED APPL BIOTECHNOL. The journal is indexed in ISI Thomson Reuters, Scopus, AGRIS, CAB International Abstracts, Biosys, Latindex,

It is intended for the publication of original scientific articles that can contribute to the scientific and technological   development of plant breeding and agriculture. The articles should cover basic and applied research in perennial and annual plant breeding, in the areas of genetics, germplasm conservation, biotechnology, genomics, cytogenetics, experimental statistics, seeds, food quality, biotic and abiotic stress, and related areas. The article must be unpublished and may not be submitted to another journal. The opinions and concepts expressed are the exclusive
responsibility of the authors and do not necessarily reflect the ideas of the Editorial Office. However, the Editorial Board reserves the right to suggest or request any necessary changes. The journal uses the Ithenticate system to identify plagiarism. Full or partial reproduction of articles is permitted, provided the source is cited. All the content of the journal, except where identified, is licensed under Creative Commons. All articles are published free of charge. CBAB is an open access journal.

Article

CBAB publishes articles exclusively in English. Authors must submit their articles in English. If the article is approved, it must be revised in linguistic terms, and the revision will be carried out exclusively by CBAB's official translators, at the author's expense. Contributions are submitted via the WEB at **http://www.sbmp.org.br/c bab/siscbab/index.php**, by clicking on **Submission**. The article management system will ask for the corresponding author's e-mail address and generate a password. **Manuscripts must be entered without the authors' names and addresses, which must be provided on a separate form**. As CBAB operates with double-blind peer review, authors should not reveal their identities in the manuscript. Authors can use their e-mail address and personal password to track the progress of their article. The article will be assessed by specialist *ad hoc* reviewers to help the Editorial Board make the final decision to accept, modify or reject the article. The complete article should preferably contain the following sequence: title, abstract, key words, introduction, material and methods,
results and discussion, acknowledgements, references, and tables and figures. It should be typed in Word for Windows, in Times New Roman font, size 12, double-spaced, A4 format, with 20 mm margins and consecutive pagination at the top right. The article should not exceed 18 pages, including tables and figures typed on separate pages (one per page) at the end of the text. All equations, models and symbols must be entered via Microsoft Equation. The title should be clear, concise and reflect the essence of the article. It should be written in capital letters and set to the left, and should not contain more than 15 words typed in bold. The Abstract should not exceed 150 words. A maximum of five keywords, other than the title, is allowed. The introduction should include a brief literature review on the topic and the objectives of the research. The Material and Method should be written in such a way that another researcher could repeat the experiment   ,

Results and Discussion should be presented together for greater readability.
Acknowledgements should be brief and limited to actual collaborators and
funding agencies.

Watch out for references.

Do not quote abstracts from events, theses or unpublished articles. These
precautions will give the article and the journal greater credibility. Citations
made in the text by the author's surname and year (e.g. Liu 1998, Pereira and
Amaral Júnior 2001, William et al. 1990) should be ordered alphabetically in
the References section, following the examples below:

Articles in periodicals:

Pereira MG and Amaral Júnior AT (2001) Estimation of genetic components
in popcorn based on the nested design. **Crop Breeding and Applied
Biotechnology 1**: 3-10.

Book:

Ramalho MAP, Ferreira DF and Oliveira AC (2000) **Experimentação em
genética e melhoramento de plantas**. Editora UFLA, Lavras, 326p.

Book chapter:

Sakiyama NS, Pereira AA and Zambolim L (1999) Improvement of Arabica
coffee. In Borém A (ed.) **Melhoramento de espécies cultivadas**. Editora
UFV, Viçosa, p. 189-204.

Congress:

Frey KJ (1992) Plant breeding perspectives for the 1990s. In Stalker HT and
Murphy JP (eds.) **Proceedings of the Symposium on Plant Breeding in
the 1990s**. CAB, Wallingford, p. 113.

CBAB also publishes other

all submitted to *ad hoc* reviewers in the same way as the articles.

Reviews

Reviews, also limited to 18 typed pages, will be requested by the Editor from
authors who are well-established in the research surrounding the topic of the
review. They will be prepared with the aim of shedding light on an exciting
topic that deserves an in-depth analysis of its state-of-the-art.

Notes

Notes are limited to 12 typed pages and are intended to report new research
or observations for which the analytical tools do not apply. They can focus on
a topic of broad interest; a short report on original research; a report on
participatory research; observations of special interest in the areas of
research, teaching, extension; the launch of new software related to the area
of improvement.

Improvement programmes

Improvement programmes
innovative or that stand out for their efficiency, impact and/or continuity may
be portrayed in the **CBAB,** limited to 18 typed pages.

Launch of   cultivars

New cultivars deserve a special section because of the importance they represent for breeding and, consequently, for national agriculture. The Launch of new cultivars section should contain an abstract, limited to 50 words, keywords, an introduction, the breeding methods used and the characteristics   of the new cultivars.
performance, basic seed production and a minimum of references, tables and figures. The entire text will be limited to 12 typed pages.

Book review

This new section has been created to announce new books related to plant breeding. Contributions to this section will be made by the author sending two copies of the book. The book will be sent to a specialised reviewer, chosen by the Editor, to write the review.

Points of view

Viewpoints, as well as reviews, will be prepared for **CBAB at the** invitation of the Editor, to cover topics of interest to breeders and society.

Letters

Short letters, also of general interest, will be accepted for publication. The Editorial Office reserves the right to edit the letters due to space limitations and clarity of exposition.

Printed by Books on Demand GmbH, Norderstedt / Germany